RV Electrical Systems

RV
ELECTRICAL SYSTEMS

And How They InterConnect

BY DONALD W. BOBBITT

ISBN - 9781699671450

Don Bobbitt

Table of Contents

RV Safety & Reality Disclaimer 4
An RV Electrical Systems Overview 5
    AC Power Systems 6
    Motorhome Power Generators 6
    DC Power Systems 6
    Motorhomes, Diesel or Gasoline 7
    RV SOLAR POWER Systems 8
    Off the Grid RV Solar Power 8
    Special Power Sources 9
The Evolution of RV Electrical Systems 11
    Luxury Equipment 12
    My Hot Water Heater Experience 12
How are RV's wired? 14
The two major RV Electrical Systems 17
    12-VDC Power Source 18
    Battery Configurations 18
    Battery Banks 18
    RV Battery Voltages 21
    Notes about Charging Batteries 21
    RV Battery Preventive Maintenance 22
    12-VDC Safety Fuses 24
    Automotive Fuse Values: 24
    110-VAC POWER 28
Campground Power Standardization 31
    Basic Power Wiring Standards 31
    A Standardized Campsite Power Box 33
    Higher Current Demands 34
    Surge Protectors 36
Modern RV AC & DC Power Systems 38
    Why have 220-VAC in an RV? 39
    AC Power Campsites and Power Cables 40
    POWER Control Systems 41

RV Electrical Systems

DC Power Control Systems 42

Managing AC Power System Overloads 43
Personal Appliances 44

Standardized Electrical Functionality 46

RV's with built-In AC Power Generators 47
The RV Generator and the Fuel Line Secret 47
SAFELY Switching between Shore and Generator Power 48
RV Generator Loading 48
AC Power Transfer Switch 49

INVERTER, CONVERTER & GFCI Functionality 51
Why have a Converter? 52
Why have an Inverter? 53
Why have GFCI Receptacles 54
Cheater Power Cords and Home Receptacles 55

COACH and ENGINE DC Power Systems 57
The CUT-OFF Switches 58
Exterior Electrical Equipment 59

The RV Propane System 63
2-Way Fridge 63
Propane Stove (Range) /Oven 65
Propane Water Heater 65
Propane Furnace 66

A Home Fridge in an RV 66

Towing Connections 68
4-Pin Tow Connector 68
6-Pin Tow Connector 69
7-Pin Tow Connector 70

Don Bobbitt

RV Safety & Reality Disclaimer

Please notice the picture I placed on the front page of this book. While camping in the Keys, not too long ago, I took the picture you see, on the day after the motorhome it caught on fire in its campsite.

It seems that the owner was "do it yourself" guy and he after purchasing the old Class-B motorhome you see he had spent many hours redesigning and upgrading the RVs wiring systems to include all of the latest technologies himself.

He and his spouse drove their "upgraded" motorhome down to the Keys for a short vacation.

Well, they were able to spend three nights in their RV before it caught fire in the middle of the night. He and his wife were able to safely escape the flames but their motorhome, as you can guess from the picture, was a total loss.

I'm showing this to my readers, because I want to emphasize to everyone that RVs are wired with built-in safety systems (breakers, fuses, and automatic alarms) that are all designed to protect the RV user and their camper.

The drawings and documentation provided in this book are designed to help RV owners figure out the possible cause of their RV's electrical problems.

In no way should this book be used by anyone to modify any RVs wiring systems or bypass any existing safety features.

Nor is the information in this book designed for any unskilled owner/user to perform trouble-shooting or electrical/mechanical repairs to their RV systems themselves if they are untrained.

'NUFF SAID!

RV Electrical Systems

An RV Electrical Systems Overview

Campers or RVs (Recreation Vehicles) are designed for the users comfort and convenience as they travel and camp.

To do this, the RV manufacturers provide one or more of several types of electric **POWER SYSTEMS** in their campers in order to provide electric power to the many built-in and plug-in appliances and equipment that make the camping experience more enjoyable for the owner.

The electrical systems that could be found in today's RVs could include;

- AC Electrical Power,
- DC Electrical Power,
- SOLAR Electrical Power,
- Propane Gas Power.

You will find that at least one or more of these Power systems have been designed into the different RV models produced over the past century.

Which RV models might have which specific electrical system designed into it, is actually a result of two things;

- The age of the RV, because over time new options were added to newer RVs, and
- The size/cost of the RV, because the larger RVs cost more and the owners demanded that more luxury items be included in their RV.

AC Power Systems

An **RV,** specifically a motorhome model, may have one or often two different AC Power systems, which provide the necessary electricity for much of a camper's electrical equipment.

For all campers, regardless of model, size or cost, the main source of **AC Power** is usually provided by a standardized **external AC power source;** which is most often a campsite's power box and is also similar to a typical home's electrical power system.

Motorhome Power Generators

In addition to an eternal Power connection, many of the motorhome type RVs include a built-in generator that could provide a second **AC-Voltage Source** as an alternative for powering the motorhome's equipment.

When a generator is built into a motorhome, the RV owners have the added convenience of AC-Power when they are traveling and camping, especially "boon-docking" or "rough camping" in the Wild.

This also includes other camping situations such as when you camp in a location where there is no external power available when you want it, such as; parking lots, Rest Stops, as well as crude campsites in the Wild.

DC Power Systems

RVs will also have at least one DC Power system, which is typically provided by one or more rechargeable batteries that are connected together to provide a steady 12-VDC system that is used to power certain internal equipment common to most RVs.

In a motorhome there are two distinct 12-Volt power systems. These systems are usually referred to as either the

- **MAIN** (or **Engine**) battery system, or the
- **AUX** (or **COACH**) battery system.

It should be noted that even the other RV designs, which are not motorhomes will have interior electrical systems and equipment that are often just like the AUX system described here, but there will obviously be no MAIN system.

There are only slight wiring differences which are necessary to provide power to such things as the exterior running lights, power steps, door light and power awning.

So essentially, camper trailers are wired the same as the **AUX** systems of motorhomes.

Motorhomes, Diesel or Gasoline

In a gasoline powered motorhome, the **MAIN** electrical system is essentially identical to that of a truck manufacturer's standard gasoline powered truck Engine system; with Ford, Chevrolet and Mercedes being the most common models used.

With Diesel motorhomes, the most common **MAIN** engine systems manufacturers are; Caterpillar, Cummins, and again Mercedes.

It should be noted that diesel motors are much higher compression than the typical gasoline engines.

This means that it takes a lot more current capacity to start a diesel motor than a gasoline one, especially in colder weather.

Because of this difference, you will find diesel motorhomes will have larger MAIN battery systems (ie. more batteries).

RV SOLAR POWER Systems

More and more often, new RV's will have some level of power provided by **Solar Power systems**.

These Solar Systems can be as small as one that provides just enough electrical power to keep the RV's batteries charged while the RV is in storage (while just sitting outside, of course) where they have access to the Sunlight.

Off the Grid RV Solar Power

Because a modern RV is already a nearly self-contained home, some RV owners will take that next step to making their RV one that can operate totally **Off The Grid.**

This is done by installing a Solar Power System that can provide enough power for their RV to operate for long periods of time, without any external power source.

These custom Solar systems can cost $25K to $50K (or more?) depending on what level of complexity is desired by the RV owner that must be designed into a custom RV Solar Power system.

These larger and more complex Solar Power systems, installed in an RV can often provide the capability for the RV owner to "live off the Grid" in their RV for weeks, even months, and still have the luxuries they desire right at their fingertips.

RV Electrical Systems

Special Power Sources

And just to confuse things for many RV owners, nearly all Campers, even motorhomes will utilize devices such as **CONVERTERS and INVERTERS**.

Converter -

A CONVERTER will utilize the 110-VAC Power Source and convert it to a useable 12-VDC source that is used to keep the RVs COACH batteries charged.

The Converter is designed to provide a *limited amount of DC power* to the COACH batteries so they can, in turn, provide DC power to certain built-in appliances and accessories in the RV.

> **NOTE:** The Converter may power some of the DC devices in an RV; but it **is not designed** to support every DC powered device in a motorhome or travel trailer.

Inverter -

At the same time, many RV's will also have one or more INVERTERS installed in them.

These Inverters are used in RVs to utilize the existing COACH battery's 12-VDC and convert it to a useable 110-VAC source for use by certain electrical devices or appliances in the RV.

Inverters became necessary when RV owners wanted to use devices such as a TV or their personal computer while the RV is not connected to outside power.

Don Bobbitt

Inverters only provide a limited amount of 110-VAC current for the RV owners so they can enjoy certain luxury devices and the owner should take care not to overload the special receptacles provided in the RV.

And, always remember that these Inverters are drawing their power form your COACH batteries and the owner should always make sure they have good, fully charged batteries in their RV before they go on the road.

The Evolution of RV Electrical Systems

Before we get into the actual electrical functions that are commonly found in today's RVs, it is good to be aware of just how far RV's have evolved over the past century.

So, here is a very short overview of how RV's, in general, evolved to the present day RV's that provide so many luxuries for the owner.

In the early 20th Century, the common camper would have very few electrical capabilities designed into it at all and what you did find installed in them was very simple.

In those times, there were really only three kinds of camping capability available to the public; either you used a tent, which eventually evolved into what is called a Pop-Up Camper; or you could purchase a hard body towable trailer which was designed to provide basic Bedroom, Living Room, Kitchen and even limited Bathroom facilities and comforts.

The camper trailer owner would often have an automotive style 12-VDC battery installed in it which was used to provide power for a few electrical appliances, such as interior lights.

Usually, the battery would be mounted outside the camper, often on the front hitch frame. This was done to keep the battery gases from leaking inside the camper itself and also for easy access when the battery needed to be connected to a charger.

Eventually camper manufacturers would add a couple of AC (110-VAC) receptacles on the inside, which were in turn wired to an external 110-VAC connector mounted on the camper's outside wall of the camper.

With this added design feature, owners could quickly hook up to, or unhook from, a campsite's AC-Voltage power source as they traveled around the country.

Luxury Equipment

Early campers were considered luxurious as compared to a tent, especially considering they would have such comfortable furniture in them as, one or more beds, a sofa, a dining table and even a kitchen sink, all built into the camper for the owner and his/her family to enjoy.

Today, when you visit an RV dealer's lot you will be amazed by all of the luxury devices; which are built into today's camper if you consider what was in one decades ago.

But, along with these added electrical luxuries, technical problems often occurred. It seemed, one thing that comes with the added use of these many luxury devices was the reality that there will eventually be more problems to overcome, and yes, repair.

My Hot Water Heater Experience

Here is an example of what happened to me one time when I had rushed through my campsite setup.

A few years ago my family and I arrived in our RV at the campground where we would be camping for a week. I was in a hurry and I quickly hooked everything up to our camper and once that was done my wife and I took our kids for a walk around the campground.

Eventually, we returned to our camper and I sat in my favorite outdoor lounge chair, just for a moment, absorbing the great atmosphere of the campground and thinking about what we would be doing over the next week.

It was only a couple of minutes later when my wife called me back inside the camper. It seems we had a problem; she had no HOT Water in our RV's kitchen.

Confidently, I tried the switch, then I looked for a blown fuse, and then I even checked for a loose wire to the Hot Water Heater. Desperate for a fix, I went over and knocked on the door of my neighbor camper. He was an older gentleman who it turned out had been camping for years and when I explained my problem, he just smiled and came over to my camper with me.

He looked at my camper, checked the power switch, and then he smiled and asked me if I had the water to the camper hooked up and turned ON.

I remembered hooking up the water line, but I was in a hurry when we arrived in the campground and I hadn't yet checked out my campsite setup to make sure everything worked properly.

And as I thought about it, I realized I hadn't remembered to turn on the water at the campsite and then flush the air out of my camper's water lines.

Embarrassed, I ran outside and sure enough, the water was connected, but I had not turned it on. I still tell my fellow campers about this instance of when my lack of attention to the numerous things that a person needs to do when connecting a camper to a campsite had stumped me.

How are RV's wired?

One thing that I have learned during my many years of camping is that most RV owners, especially the "Newbies" have no idea how their campers are wired or how their electrical systems operate.

The typical RV owner will normally just pull into their campsite and plug their RV into their campsite power box and expect everything electrical to immediately work for them.

And until there is actually a problem they won't even think about where the power that makes everything work in their RV comes from or how it is managed.

So when a problem does occur with something in their RV, the first thing an RV owner looks for is a;

- **schematic of the actual wiring**, or maybe a
- **trouble-shooting chart** on how to find the cause of the strange problem they are having with their RV.

This is usually the point where the RV owner, especially the Newbie, finds out that no one has actually ever pulled this kind of information together into a single document which they can quickly use to find and fix their RV electrical problem.

Of course, the numerous manufacturers of RVs will have their own Engineers and each company Engineer will design their RVs wiring to suit their specific needs.

So, they do have wiring diagrams and technical documents that they use internally, but they do not make these electrical details available to the public. They do not share

RV Electrical Systems

this level of documentation to the average RV owner for a number of their own reasons, mostly for the safety of the average non-technical RV owner.

On the bright side of this situation, while the manufacturers may not wire their RVs exactly the same, they do tend to wire them to function the same way as their competitors RVs.

And much of the commonality in RV designs comes from the fact that the major RV manufacturers will often use the same commercial sources for so many items installed in their RV, such as;

Hot Water Heaters, Water Pumps, Lights and fixtures, Refrigerators, Microwave Ovens, Washer/ Dryers, other Appliances, Video and Audio equipment, and more.

Because of this commonality of the electrical and mechanical parts and appliances used in almost all RVs, the designers are often forced to wire their RVs the same way.

So, even if an RV owner or the trained RV Service technicians may not have the actual schematics of an RVs electrical wiring, they will generally know just how things are put together and how they operate.

> **SAFETY NOTE**: Even if you are not a trained technician, you should at least be able to give the right data to a trained service technician so they can repair the actual problem.

This book, **Typical RV Electrical Systems**, is designed to provide the reader will a good understanding of how the electrical systems in most RVs operate and by understanding the block diagrams shown here you can see just how all of the technical equipment and systems are inter-connected.

So, read on and you should not only quickly understand just how all of these specialty devices and electrical systems are interconnected, but you should also understand better how to troubleshoot almost any electrical problem down to the device that is causing your problem.

I repeat, this is not a collection of specific schematics of the wiring in any manufacturers camper electrical systems, but rather it is **a collection of GENERIC ELECTRICAL Functional diagrams**.

And as I have said, these Generic Diagrams should help any camper owner understand how the many electrical devices in a camper are connected and help them understand how they function and often even find the source of their problem.

Remember that the drawings you see here are not specific wiring schematics of any model or make of camper and they should be used as what they are, functional diagrams of the more popular RV electrical systems built into RVs.

RV Electrical Systems

The two major RV Electrical Systems

There are at least two major electrical systems built into all RVs made in the USA.

If you look at the diagram below that is labeled; **Simplistic Generic RV AC & DC Power Systems - dwg-01**, you will see a simple diagram of the two major electrical systems that you could find in almost all campers and motorhomes in the past.

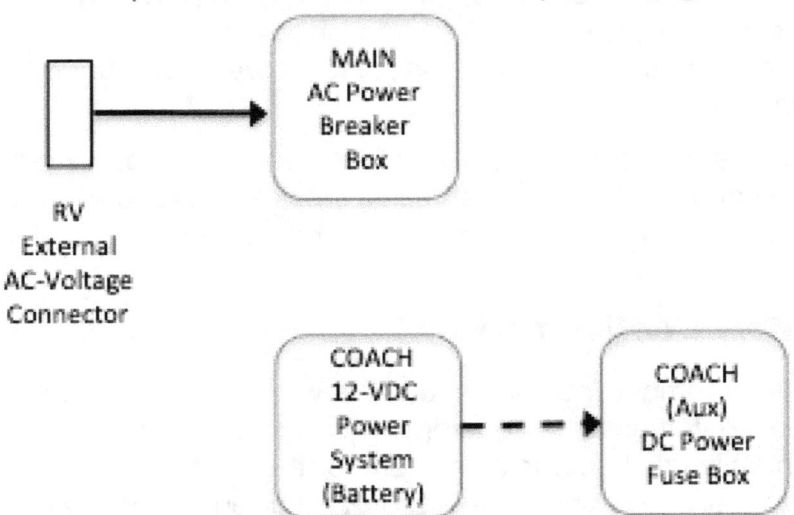

The drawing is very simple and the reason I am displaying it here is to illustrate how simple the electrical wiring in the original campers, built during the early-20th century were.

Even though modern RV electrical systems are far more complex, you can still find this kind of basic electrical system in a simple Pop-Up Camper as well as those very old camper trailers.

12-VDC Power Source

The very early designs of manufactured campers and travel trailers would have an automotive grade "lead-acid" battery installed in it.

This battery was used to provide a 12-VDC Power source for the few interior items that were built into the early trailer campers and they still are today in some models.

This electrical system is often referred to as the; **AUX** (or **COACH)** DC Voltage system.

At one time, the appliances and accessories powered by this system were such simple appliances as; interior lights, a 12-Volt Fan, and maybe a small electric water pump for the kitchen sink.

These items typically used very little power so people could often camp for days, sometimes a week or more, using the same battery before they needed to recharge it.

Battery Configurations

On later camper models the owners wanted to use even more 12-Volt electrical equipment while using it, and eventually this became a problem for them when using only one battery to support these growing current demands on their 12-VDC Power System.

Battery Banks

You will often find that you have multiple batteries wired together into a "bank of batteries" in order to provide 12-

VDC with an even higher current storage capacity for the RV.

These battery banks could be configured using either one of more 12-VDC batteries as seen in **Fig.DC-01 Multiple 12-VDC Battery connections**.

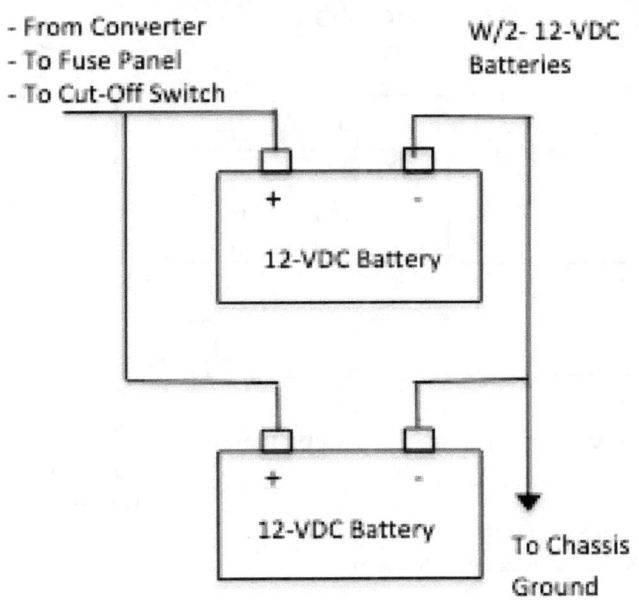

Fig. DC-01 Multiple 12-V Battery connections

The COACH batteries in a Camper are almost always "deep Discharge" designs because they are constantly being loaded, partially drained of power and recharged.

By using the popular **12-VDC automotive deep-discharge batteries,** the RV owner can more easily find the specific battery design replacement they need; because these deep discharge batteries are popular around the country for use in **Marine** applications and can be found at so many automotive parts stores.

Also, the wiring of multiple 12-VDC batteries together, in parallel, is simpler to implement, than with multiple 6-VDC batteries.

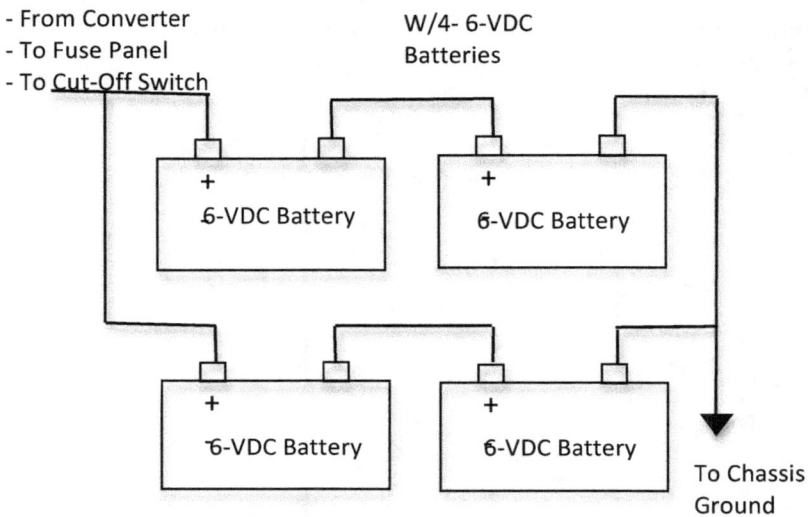

Fig. DC-02 Multiple 6-V Battery connections for COACH 12-VDC System

An even higher current capacity can be provided by using combinations of two, four, six or more 6-VDC batteries, as shown in **Fig.DC_02 Multiple 6-VDC Battery connections for a 12-VDC system**.

Because of the number of larger "plates" inside these 6-VDC batteries you can get this increased Current storage capacity for similar sized battery packages.

In either case, these different voltage batteries would be configured to provide the same 12-VDC to your camper's DC accessories.

RV Electrical Systems

RV Battery Voltages

The table below is a great one to use when you are concerned about whether your RVs batteries are properly charged or not.

You should use a digital Multimeter and measure the voltage directly across the Positive and Negative terminals of the battery you are concerned about.

Of course your battery may show any voltage level in the range of those shown in this table, but you can use these numbers to figure roughly where your batteries are.

Battery Charge Levels

12-Volt battery	6-Volt battery	Percent Charge
12.65 Volts	6.3 Volts	100%
12.45 Volts	6.2 Volts	75 %
12.24 Volts	6.1 Volts	50 %
12.06 Volts	6.0 Volts	25 %
11.89 Volts	6.0 Volts	Discharged

Notes about Charging Batteries

While discussing RV batteries and recharging them, you should understand these common facts about them.

1. After fully charging a battery, wait at least 12 hours to check the actual open-load voltage. This wait will allow the plates of the battery to stabilize and will provide you with a more accurate reading.

2. **CHARGERS**: Almost all RV's today, like automobiles, utilize numerous computers in their electrical systems, and it

is not recommended to use fast-charge "jumper chargers" or "booster packs" to start an RV or car quickly, as the higher voltages used can damage the RVs computers.

If your batteries are maintained and used properly, you should never need a jump; but if you do, it is recommended that you utilize a standard slow charger to avoid damage to the batteries and the RVs electronics.

RV Battery Preventive Maintenance

Remember, you count on your RVs batteries every day, whether camping or traveling, and all of your batteries require **regular preventative maintenance**.

Like many of the "Full-Timers", and even "Part-Timers" I know, I recommend that you perform the following preventive maintenance steps, as a minimum, on a regular basis:

- **Inspect Connections.** Inspect your batteries and the whole battery compartment along with the wiring that is visible.

    Remember your RV is bouncing over bumps and swaying around curves. Always perform a good and thorough visual inspection of your batteries before a trip.

- **Inspect for Damage.** After you set up at your campsite, you can save you a lot of heartache later if you check for such things as loose wires, frayed wires, a cracked battery case, spilled battery fluid, or battery acid build-up on your terminals that may have been caused by your travels.

RV Electrical Systems

- **Battery Fluid Level.** Before every trip, **check all battery fluid levels** and refill any cells that are low on fluid as necessary, with **distilled water** only.

- **Road Ready.** Before every trip, check that all battery connectors are tight, and are clean of any oxidation residue.

- **Storage Inspection.** When storing your RV for an extended time, check the batteries at least once a month and either utilize a small Solar Trickle Charger or run your generator until they are fully charged. See the Table below for relative battery capacity.

- **Residual Current Leakage.** Remember, these electrical systems, even with the switches turned off, will leak low currents and this will, over time, discharge your batteries.

- **Inspections While Camping.** When camping at a campground for an extended time, check the batteries at least every 2-4 weeks, especially the fluid levels in the coach batteries. Remember, these batteries are powering most of your lights and many appliance control systems, so even though you are plugged into 110-Volts AC at your site, they are constantly being re-charged and can lose fluids over time.

- **Start Engine Monthly.** Also, when camping for an extended time, start your coach engine at least every 30 days, and run it for at least 30 minutes to keep every thing lubricated in the engine system, but also to re-charge your chassis batteries.

12-VDC Safety Fuses

With all of these electrical devices that demanded 12-VDC Power, the manufacturers added a 12-VDC Fuse Box that was connected to the COACH battery bank.

Early RV Fuse panels would use the old glass-cylinder design of fuses, but a couple of decades ago, they evolved over to using the standard automotive fuse.

These DC-Voltage fuses are color-coded and they are also marked with the actual fuses current limit.

The **COACH DC Fuse Box** usually includes one or more 30-AMP fuses which are on the INPUT lines from the battery to the Fuse box.

Typically, these high current fuses should not be blown because at least one or more of the other lower value circuit fuses should have been blown first if the circuit they are protecting has a problem.

The rest of the fuses in the Fuse Box are for protecting individual DC circuits wired into the RV and they can vary in the value needed for each circuit.

Automotive Fuse Values:

Automotive fuses are simple devices that will "blow" when there is too much current being drawn through them, even if for only a millisecond to a second.

They are designed to operate at certain specific current limits and the user can determine the value of one of these

RV Electrical Systems

fuses by either the marking on the fuse or the color of the fuses plastic body.

See the table below for a listing of these standard fuse values. And you may want to purchase spares in case you do blow one while traveling.

Standard Automotive Fuse Values and Colors	
Fuse Value	**Fuse Body Color**
1 Amp	Black
2 Amp	Gray
3 Amp	Violet
4 Amp	Pink
5 Amp	Gold
7.5 Amp	Brown
10 Amp	Red
15 Amp	Blue
20 Amp	Yellow
25 Amp	Clear
30 Amp	Green

The other fuses protect the circuits that provide the needed power for certain common electrical systems such as; Interior Lights, Alarms, 2-Way Fridge, Temperature Control Panel and the Power Control panel.

The fuse panel for the COACH 12-VDC electrical system was more a safety upgrade than anything for RVs; because it in turn provided the owner with separate electric circuits that were previously not managed.

The DC fuses protected the camper equipment from potential damage that might occur when there was an overload condition or if a direct short occurred on one of the built-in DC circuits.

Over the years, other equipment that could operate on 12-VDC has been installed into even the simplest campers for the owner's convenience and safety, such as;

- Smoke Alarms,
- Carbon Monoxide (CO) Alarms,
- Exterior camper lights,
- Small 12-VDC fans
- As well as certain other luxury items that used very little current and could operate on the 12-VDC provided by these batteries.

Where do you find your Fuse Boxes?

I find that this is a question asked often by the Newbie RV owners.

The problem is that over the years the RV manufacturers placed them wherever it was convenient for them and not necessarily for the RV owner.

From experience I can give you a few suggestions of where to look for these elusive fuse boxes;

For a motorhome, the Engines Main fuse box is going to be under the hood the same as it would be with a normal truck of the same brand.

Dash accessories that are powered by the engine electric system will usually be connected to a fuse box mounted most often under the dash and on the driver's side.

COACH equipment and accessories will also be connected to a fuse box, but its location can be harder to find.

RV Electrical Systems

Depending on the model of the RV and the year it was manufactured, I have found that most Coach Fuse boxes or panels could be;

- Over the Kitchen range,
- In the Bedroom, on the wall opposite the Bed
- Near the Entrance door behind a panel on the wall,
- Behind an inside panel near the Coach battery location,
- On the small panel under the Fridge,
- And I'm sure there are other hidden locations I haven't come across myself.

Recharging Batteries: Remember though, these automotive batteries in your RV can only provide a limited amount of Current before needing to be recharged.

Because of this the camper owners would only have this 12-VDC power for a few days of use.

If there wasn't some way to recharge them the trailer owner had to take the battery somewhere that there was a charger to recharge their battery.

And, because these COACH batteries had to be recharged over and over, they had to be a "**deep-discharge**" marine-type design that could handle the constant cycle of being discharged and recharged with out being damaged.

As mentioned, the RV's built-in **CONVERTER** operates on 110-VAC power and it will keep the Coach battery charged as long as the 110-VAC Power was available to power the Converter.

110-VAC POWER

Quite early in the evolution of these basic camper trailers many of the potential customers for them began demanding that there be access to 110-VAC in their campers.

The camper manufacturers quickly responded by adding 110-VAC receptacles that were built into the interior of the camper so the owner could use their favorite home appliances while traveling and camping.

See the diagram **Simplified RV Interior AC-Voltage System & Equipment - dwg-02a** for a simple diagram of an RVs AC Power system.

The equipment shown is what you will find in a typical Camper Trailer.

A motorhome, which has a more complex AC-Voltage Power system, which includes a built-in power generator, has a significantly different configuration.

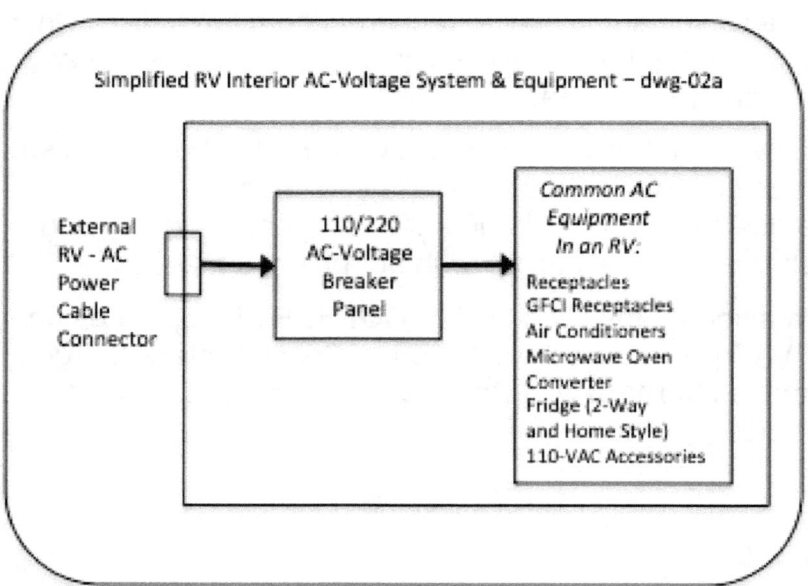

RV Electrical Systems

RV manufacturers standardized their interior 110-VAC power system using common electrical breakers that were wired into a breaker box (or panel) that allowed the safe control of this AC-Voltage power system for the RV.

The addition of the AC breakers in a breaker box also added a level of power control and safety for the RV Owner.

And all of the RV manufacturers quickly began to copy the wiring standards already in existence and used across the country.

This standard wiring design concept was also used in almost all of campgrounds around the country.

Even the electrical connections from the different campground's electrical power boxes that were eventually used at every campsite were standardized

To connect to a campsites power box provided by campgrounds these original RV 110-VAC systems all utilized a standardized and waterproof power cable that was capable of handling the voltages and current provided by the campsite.

It was soon evident that whatever was used to connect an RV to a power source in any campsite had to be the same standard design so that any RV owner could use the power provided.

Once there was such a standardized External Power cable available, the RV owner could go almost anywhere and purchase one and then they could easily connect their camper to a standardized campsite power box provided at each campsite.

These standard RV power cables would have a female connector on the RV end and a male connector on the power source end.

Once they had one of these cables for use with their RV, the camper owner could travel to their favorite campground, pretty much anywhere around the USA.

Then they could plug their camper into a **Campsite Power Box** and have access to the 110-VAC/220-VAC power they needed to use while camping.

Camper owners were then able to use their personal electrical appliances and accessories in their camper.

Such appliances as; *fans, radios, table lamps, Coffee Pots, Portable heaters, Hair Dryers*, as well as other such personal electrical equipment that people would use in their daily lives became common in RVs.

RV Electrical Systems

Campground Power Standardization

Once campers could use their new external **standardized** electrical power source, this forced the campgrounds to come up with their own connection system for their campsites that not only provided power, but also protected their campground's electrical systems.

With the addition of a Campsite Power Box, as long as the camper didn't overload their campsite's standard electrical breaker system, which was enclosed in the Campsite Power Box, all was well.

Once the two connection ends; the campsite and the camper, were defined, the wires to the connector ends had to always be the same.

Basic Power Wiring Standards

All electrical wiring used in a home or motorhome must be wired according to electrical codes and utilize the wiring definitions listed here and must be wired consistently to assure the safety of the user.

There are three necessary connections that must be in place and they are;

GROUND – The Ground wire is connected to all other ground wiring and can be traced electrically back to an actual connection on a rod into the ground itself.

COMMON – This is not the came as the Ground, and it is considered to be the electrical return path for the electricity in a device.

HOT WIRE – The Hot wire is the wire that supplies the power for the device it is connected to. It is the source of the Voltage and should only be handled by a qualified electrical technician.

When these three wires are all connected properly then the device should operate properly and safely.

See the diagram; **RV AC-Voltage Wire definitions - dwg-AC-01,** which attempts to show how the different wiring systems were designed for RVs.

RV AC-Voltage Wire definitions – dwg-AC-01

110-VAC
15-20-Amp
3-wire Service

> 110-VAC - Hot
> 110-VAC – Common Wire
> 110-VAC – Ground Wire

110-VAC
30-Amp
3-wire

> 110-VAC – Hot Wire #1
> 110-VAC – Common Wire
> 110-VAC – Ground Wire

220-VAC
50-Amp
4-wire

> 110-VAC – Hot Wire #1
> 110-VAC – Hot Wire #2
> 220-VAC – Common Wire
> 220-VAC – Ground Wire

You should use this drawing; **(dwg-AC-01)** to see the definition of what specific connections each of the wires provides when you connect your RV to your campsite power box.

Originally, the campground power boxes were only wired to provide the RV owner with **15-Amps** or **20-Amps** of 110-VAC electrical service for a single campsite.

RV Electrical Systems

Over time, the demand for more and more power in RVs forced the manufacturers to wire their campers to provide for this growing need.

Eventually they evolved their campground wiring to a standard **30-Amp 110-VAC** service, which could then power certain higher current appliances such as roof air conditioners and clothes dryers.

But even with this new level of electrical power in their RVs, the demand for more power kept growing and eventually the camper power system market settled out with what is now a standard **50-Amp 220-VAC** Electrical service.

This 50-Amp service is essentially a combination of two standard 20-AMP sources that are combined in a cabling system made up of a wiring system and power cables capable of carrying this added load without having to use larger and heavier wiring.

A Standardized Campsite Power Box

The campgrounds that were popping up around the country began providing an **onsite power box** that was wired to popular local electrical standards.

And as mentioned, these new campsite power boxes provided standard electrical connections and breakers, for the convenience of the camper owner as well as for the protection of the Campground's electrical supply system,

To this end, the campsite Power boxes usually had their own circuit breakers built into them.

Once this was done their campground power source was protected from damage in case the RV using it had an electrical problem; or if it was attempting to use too much power from their campsite electrical system.

At first, the power provided at campsites was provided at a

relatively low current such as the early 15-Amp or 20-Amp 110-VAC service systems.

And because there were already standardized exterior connectors and cables on the market, the camper trailer manufacturers standardized on these special parts designs.

This input power to an RV was wired to the **Main AC Power Breaker box**, which would have a Main breaker with one or two circuit breakers in it at first.

This wiring system provided the RV owner with an external electrical system designed to handle the small loads that the RVs were wired to use.

Higher Current Demands

Originally, these 15-Amp and 20-Amp AC electrical power systems were adequate for camper owners but it only took a few years before the demand for even more camper power drove manufacturers to implement 30-Amp and then eventually 50-Amp AC-Power service into new RVs.

As these newer and higher powered RV designs became common the campgrounds were forced to react with higher power sources at their campsites.

What about the Future? So far, the campgrounds around the country have resisted demands for them to provide yet another level of AC-Power capability to their campsites.

Up until now, when you open that power box at your campsite you will generally just find the four different

RV Electrical Systems

standard electrical connectors in the box for you to use, specifically;

- 15-Amp 110-Volt Service
- 20-Amp 110-Volt Service
- 30-Amp 110-Volt Service
- 50-Amp 220-Volt Service

With the newer campers being designed to be capable of using this much power, all of the campgrounds, in conjunction with the local power companies, had to redesign their entire campsite power systems

Because of the expense of changing the (usually installed underground) wiring of a campground, many of them have taken the longer view of only upgrading a few sites at a time from having a 30-Amp service to a 50-Amp service.

Because of this upgrade expense, it is still common to find RV Campgrounds around the country that will have only 30-Amps campsites and others will have both 30-Amp and 50-Amp campsites.

The diligent RV owner should always make sure that when you make campsite reservations that you specify which electrical service you require; either 30-Amp or 50-Amp.

Today, the RV owner has so many electrical luxuries in their campers that they are often laid out nicer than many people's homes and its normal to find such functional electrical equipment in almost any RV, like;

- Multiple Digital Televisions,
- Convection Microwave Ovens,
- Multiple interior and exterior Stereo sound Systems,
- High Definition Satellite TV Receiver and Tracking Systems,
- Multiple Heat Pump style Air Conditioners,
- Electric Fireplaces,
- Home-style, full size Refrigerators,

- And even built-in RV Internet/WIFI Networks,
- to name the more commonly used today.

Surge Protectors

I have known many RV owners who have had to pay for damage done to their RV's AC-Power system because of dangerous variations in the voltage provided for campers by the campgrounds local power systems.

Because of these variations, I recommend that every RV owner who travels a lot and stays on campgrounds they are not familiar with purchase two devices.

The first thing they should purchase is a good multimeter so they can keep an eye on just what level of voltage they have coming into their RV.

He second item they should own is a good Surge Protector that can actually protect your RVs electrical systems when the power coming into the RV varies wildly giving the RV what are often destructive voltage levels.

In reality your RV already has surge protection devices: they are your main AC breakers and the individual appliance and equipment breakers in your main breaker panel.

Like surge protectors, your RVs breakers kick out if the input voltage/current goes too high.

The only real difference between breakers and a commercial "surge protection" device is that standard breakers are slow to react to voltage changes.

On the other hand, good surge protector should react much faster than a breaker does to voltage and current extremes and "kick offline" when the power supplied voltage exceeds the safe limit of your electrical equipment.

RV Electrical Systems

Because low input voltages can also harm electrical devices or make them run erratically, most surge protectors will also turn the power off when the voltage is too low.

Now the problem with surge protection devices is that there are no real legal specifications for their design and function.

> **NOTE:** Anyone who buys a Surge Protector should make sure they get one that has a relatively fast response time.
>
> Its this fast response time in controlling a Power Surge that can save your RVs electrical equipment from damage.

Modern RV AC & DC Power Systems

As these original electrical Power systems in campers evolved and became more complex; at the same time the Campground and RV manufacturer industries started standardizing their electrical products, components used, and even the systems used in the RVs.

To illustrate a typical modern RV electrical system, the diagram seen here labeled; **Generic RV AC & DC Power Functional Diagram-02** shows a more detailed view of an RV's common AC and DC Power systems.

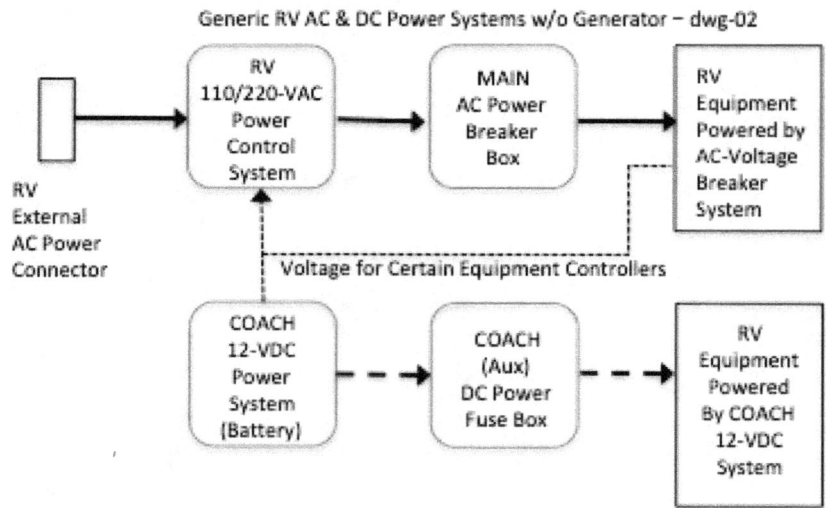

Why have 220-VAC in an RV?

One of the popular new electrical accessories that were being installed into campers at this time was an array of different manufacturer's Air Conditioners, Clothes Dryers and other different electrical devices that use a lot of electrical Current.

The Air Conditioners were usually mounted on the RVs roof for better air flow inside the RV.

Remember your Physics class and the basic statement that Hot Air is thinner than Cold Air so Hot Air will Rise while Cold Air will drop. So, by placing an Air Conditioner onto the roof of an RV allows the Cold air to drop down naturally.

Air Conditioner manufacturers soon realized that the size of the large RV's demanded a high capacity Air Conditioner.

Initially, the standard RV roof Air Conditioners have been designed to provide 5000 to 8000Btu cooling capacity for smaller campers and eventually up to 12,000 to 15,000 Btu cooling capacities for the larger RVs.

One immediate problem was that the higher capacity Air Conditioners (typically 13,000 Btu or 15,000 Btu), when operating on a lower capacity service could draw as much as 15-Amps when starting up and typically around 6 to 8-Amps while running, with peaks in current drawn every time the unit cycled.

This was a problem for RVs with older campers with a limited electrical service such as 15-Amp or 20-Amp RV once had.

Because of this problem, this new demand for more current drove RV manufacturers to add 220-VAC into the newer RV electrical systems, and this utilized wiring that provided for the two 110-VAC input wires to each have a current capacity of 15-Amps for use in the RV.

This same wiring philosophy applies to the newer, larger RVs that have a total of 50-Amps capacity for the convenience of the RV owners.

By using these limited capacity Air Conditioner designs the manufacturers could use just one, or two or even three Air Conditioner units. They were all mounted on the roof to provide adequate cooling throughout an RV for the convenience of the owner.

Other high current appliances that became common on RVs were equipment such as the Home-style Clothes Dryer; which would itself require a large amount of power to heat and dry those clothes.

By providing two 110-VAC lines to all RVs in the 30-Amp and 50-Amp service wiring, the wires in the cables themselves were smaller wires than would have been necessary if the designers had simply provided even higher current capacity by using much larger wires.

AC Power Campsites and Power Cables

So you might ask; Just what is the big problem with using an external power cable that can handle even more than 50-Amps for RV Power Systems?

One reality is the fact is that the cable needed for this high current service ends up being a bulky, very heavy and relatively expensive piece of equipment for the average camper owner to haul around their campsite and use, each time they would hook up.

This becomes more of a problem because the higher the current that is used with a cable connection demands the

RV Electrical Systems

use of larger and heavier copper wires in the Power cable used between the RV and the external power source.

And of course, while considering these higher capacity services for campgrounds and campers the fact that the wiring in the campgrounds and inside the RV would also have to be much larger, has been an important part of such considerations.

This size and weight problem is often a special concern for the RV owners themselves because they are the ones who will have to handle any newer and bulkier power cable every time they would hookup and disconnect from a campsite.

When considering the number of times the RV owner must haul out such a heavy cable to connect and disconnect their RV to campsite power sources wherever they go, then it is important that the external power cable must be useable by anyone who is in average physical condition.

So, essentially the braided wires used in a 50-Amp RV's external power cable is presently right at the limit of what is acceptable for people enjoying a camping lifestyle in their RV.

Remember that these are the same limiting factors for the RV manufacturers as well as for the campground industry who would have to rewire their whole campground electrical system with much more expensive electrical wire and controls for their campsites.

POWER Control Systems

As the functionality of a camper's AC-Power System was expanding; at the same time the DC-Power System was also expanding in RVs.

DC Power Control Systems

The days of only having a simple 12-VDC source to power the interior lights in an RV are long gone, and today a good 12-VDC Power System is critical to the functionality and safety of other equipment in modern RV's.

Equipment and appliances as varied as;

- the control circuit board of a 2-Way Fridge,
- the interior light systems,
- the Fire and CO Alarms built into all RVs today,
- the RV Power Slides
- and the interior temperature control panel (for Heating and Cooling),

Having all of these devices has driven the need for larger 12-VDC Fuse panels and larger, higher capacity deep-discharge COACH batteries; and along with them, the charging systems that are needed to supply this increased current capability.

> **SAFETY NOTE**: The sizes of the fuses used in RV circuits are designed to handle the specific maximum current that can be safely used in the circuit it protects.
>
> Because of this, you should NEVER use a higher rated fuse to replace a fuse that has blown.

Presently these DC Systems are only protected by Fuse Panels using simple Automotive style Fuses.

Managing AC Power System Overloads

Eventually the vast majority of RV's had either standard 30-Amp or 50-Amp AC-Voltage electrical systems built into them.

And it was discovered that under certain conditions there could be problems from having kicked out campsite breakers to serious safety and functional problems that were caused by the camper owners.

Far too often, they would plug too many AC-Voltage devices into the RVs AC-Voltage receptacles while also running their Air Conditioners, Clothes Dryers, microwaves and other standard equipment.

Customers would still overload electrical systems

It soon became evident that even though a camper might be designed for only 30-Amps or 50-Amps of normal power usage, there were some camper owners who would ignore the design limits and arbitrarily overload the RV's electrical system, if they were not careful.

Because of this, along with a few other potential power loading problems, the RV manufacturers installed more functionality control into the RVs that we call an **AC Power Control System**.

To put it simply, in order to handle these overload situations, and other unsafe conditions that could overload a campsite's power box capability a method of limiting this excess power usage was needed.

So the manufacturer installed certain electronic and electrical controls into the camper's wiring that could sense how much current was being used at any one time in the RV.

And if the level of power used was approaching the upper limits of the RV's AC Power system (30-Amp or 50-Amp), rather than have the campsite's breakers kick out, the AC Power Control System would begin limiting the power it provided to certain high current devices.

Once these Power Control systems were installed, the camper owner did not have to worry about using more power than their RVs electrical systems can manage.

And at the same time, the campground's electrical systems would not get overloaded and have their campsite breakers kick out

And an added advantage is that this AC Power Control System also eliminated the chance of having their campsite's power breakers "kicking out" on them under heavy load conditions.

Of course, this type of Power Management system puts the responsibility onto the RV owner to manage their AC Power usage themselves.

This is because, when the RVs internal Power Control system starts to turn OFF the power to segments of the RVs electrical equipment because the RV is drawing too much current, then the owner needs to figure out how to manage which equipment and accessories they need to use at one time.

Personal Appliances

As mentioned before, the RV owner always needs to remember that there are several built-in pieces of home-style equipment that are now common and that draw large amounts of current as compared to the older RVs.

RV Electrical Systems

And of course there are numerous personal devices that RV owners and their families bring onto their RV and use that draw a lot of current.

The use of these devices, if not managed by the owner, can, as more and more of them are plugged into the RVs receptacles, start kicking AC-Voltage breakers; and the **personal appliances** that need to be managed the most are;

- Coffee Pots,
- Clothes Irons,
- Hair Dryers and other hair management devices
- Electric Hot Plates
- Printers and PCs
- And Electric Crock Pots and Deep Fryers to name the most popular power users.

And remember, the current used by all of those; Cell Phones, device and battery chargers your family often uses can add up and add another level of aggravation to your Power management situation.

Standardized Electrical Functionality

Eventually, most of the RV's sold by the RV manufacturing industry, were wired so similarly that an RV owner or an RV Service Center technician could often, and with some confidence, understand and check out the electrical systems of almost any RV when there was a problem.

If not for this casual standardization of Electrical Systems, you can imagine that if every RV manufacturer's product was wired randomly different from the other designs built by their competitors, the world of RV Service and Maintenance would be much more chaotic than it is.

And of course, because the manufacturers were using home electrical wiring standards and equipment that not only passed local wiring codes the electrical equipment used in most RV's could more easily be repaired or replaced when necessary.

So, it turned out that an owner of one brand of RV could expect to see the vast majority of the electrical parts used in their camper would be the same as what was used on their friends RV.

And, they could expect to see the functionality of two different RV manufacturer's electrical systems and built-in equipment to operate very similarly.

RV's with built-In AC Power Generators

Many people; especially Outdoorsmen such as; Fishermen, Hunters, Nature Photographers and others who love outdoor sports were using RV campers for their travels and adventures.

The main thing they wanted in their RV though was to have an AC-Voltage power source for their camper even when they were "rough camping", which is the common name for camping where there is no 110-VAC available.

To that end, the RV manufacturers began installing certain gasoline and diesel powered generators into their larger motorhomes.

For simplicity's sake, these generators would always use the same fuel source as the motorhomes engine.

And once this added accessory was installed, the owners could now travel anywhere and still have all of their AC-Voltage powered luxuries functioning in their motorhome.

The RV Generator and the Fuel Line Secret

The addition of a built-in AC Power Generator in an RV caused a new and unforeseen problem. People were going out into the Wilds of America in their "Power Self-sufficient" RVs and using up all of their fuel.

It seems that they often did not realize they were using up the fuel in their tank without realizing it, and then they were stuck in the wilderness with no way to get back to civilization.

So, the RV manufacturers came up with a solution by only running the fuel line for the generator down to the ¼-tank level in the RVs fuel tank. This way the generator would stop running and the RV owner was guaranteed to have at least ¼-tank of fuel to get their RV back to civilization.

SAFELY Switching between Shore and Generator Power

Review the diagram labeled; Generic RV AC & DC Power Systems with Generator dwg-3 for a functional overview of how the AC and DC power systems are wired and how the generator functions as a part of the RV's electrical system.

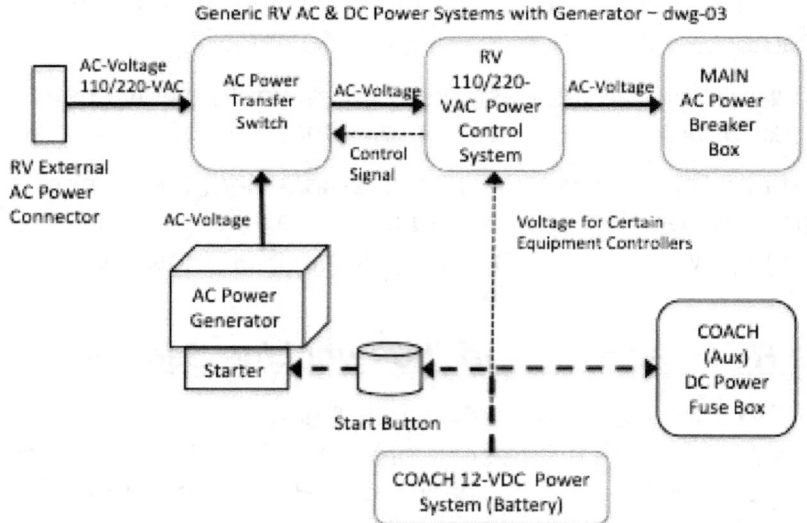

RV Generator Loading

Every motorhome owner should keep in mind that most of these generators operate best when they have a certain amount of load on them.

RV Electrical Systems

Many people will complain that their generator is "loping" or essentially speeding up and slowing down.

This typically happens because the generator doesn't have a load on it. For your RV generator to function properly then always turn a few electrical items ON and then check if it "levels out" and runs smoothly.

AC Power Transfer Switch

Once motorhomes were built with this optional alternative AC-power source, the generator, it was necessary to make sure that the two AC-Voltage systems, the Generator and the External AC-Voltage sources, were not connected at the same time.

Simply put, without going into the technical details, two **Alternating Current Power Sources** had to be synchronized to avoid serious problems that could damage the electrical equipment that operated on 110-VAC, in the RV.

To avoid this, the simplest solution for RV's was a switching system of some kind.

In order to have an RV's electrical system be capable to safely using two different AC power systems; such as an external AC-Voltage power source, and also a built-in AC-Voltage generator, was to design a way to only allow one or the other power source to be used in an RV at one time.

To this end, the RV manufacturer's engineers installed an **AC Power Transfer switch** system in all RV's that had a built-in AC Power generator.

This switch is a high current solenoid that can handle the maximum power provided by either the generator or the external power source.

The transfer solenoid, used in the Transfer Switch system, is controlled by the RVs built-in **AC Power Control System**.

Its connections are normally in the default position for using the External AC Power Source; which only allows the external power to be connected to the camper's electrical system as the default RV power source.

But, when the AC Power Control System detects that the built-in generator is running; the system applies 12-VDC power to the solenoid and the solenoid switches from the RV's external power cable and over to the generator's output as the primary Power source for the RV; thus disconnecting the External Power connections from the RV, at the same time.

> **NOTE**: This Transfer Switching System is a motorhome safety feature that guarantees only one AC power source will ever be connected to the RV's internal AC-Voltage system.

RV Electrical Systems

INVERTER, CONVERTER & GFCI Functionality

There are two pieces of electrical equipment that are installed in most RVs that many owners will typically not understand exactly how they operate.

These specialized pieces of electrical equipment are commonly called **Inverters** and **Converters**.

Generic RV Inverter & Converter Functional drawing-04

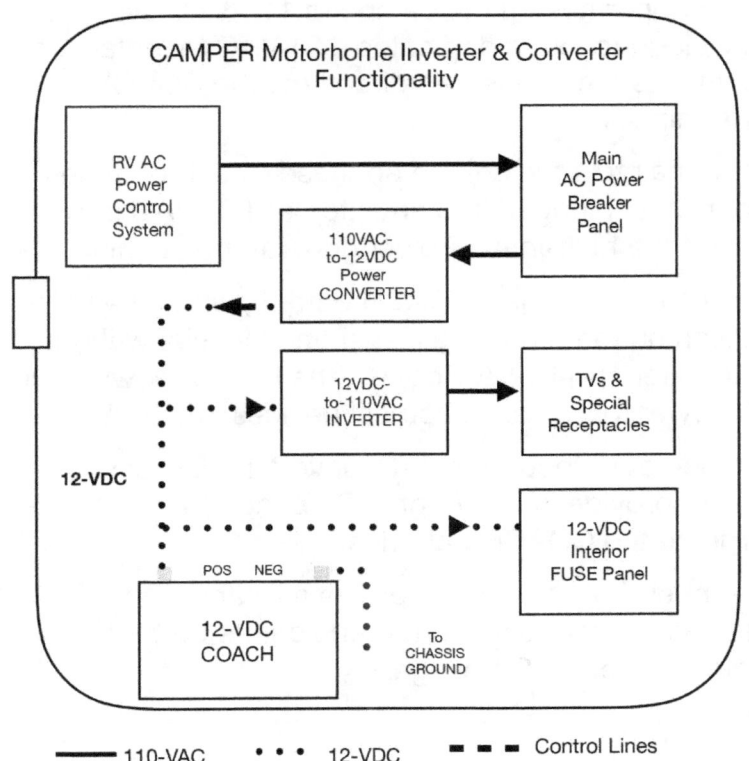

Refer to the diagram labeled; **Generic RV Inverter and Converter Functional drawing-04** for a clear view of how they are connected.

Why have a Converter?

Probably the more important of these two devices is the Converter. Although called a Converter, this device is essentially just a customized Battery Charger.

It is designed to charge and maintain the charge on the COACH batteries that you will find in almost all motorhomes and other styles of campers these days.

A Converter operates on the camper's 110-VAC system. It is usually plugged into a dedicated 110-VAC receptacle and its output wires are connected directly to the COACH battery terminals.

It exists to maintain the charge on these Coach batteries while your RV is plugged into an external 110-VAC power source, or if the built-in AC Power generator is running.

Then, with a fully charged COACH battery, when the RV is being driven on the road or is just sitting in a site without access to external 110-VAC power, the RV owner will have 12-VDC power for certain electrical devices in the RV.

It should be noted though that a Converter is usually designed to provide just enough DC Voltage to maintain a full charge on the COACH batteries.

But remember that the Converter has a limited 12-VDC charging capability and it cannot replace the battery as the RV's main source of DC Voltage.

RV Electrical Systems

Why have an Inverter?

An Inverter on the other hand, is an electrical device that can take a standard 12-VDC source, such as your Coach batteries and generate a limited amount of a simulated 110-VAC power.

RV manufacturers adopted the use of Inverters in their campers to provide the necessary 110-VAC for a few special receptacles when there was no external 110-VAC power source such as when on the road.

> **Tech-Talk Note**: The wave-shape generated by an Inverter is not a pure sinusoidal wave-shape as is in your home electric system.
>
> It actually generates a modified square-wave shape that is close enough to be an acceptable 110-VAC source for most electrical appliances such as TVs, Chargers, etcetera.

By connecting to these receptacles the RV owner could provide 110-VAC for their Televisions and their personal computers, even when their camper was not plugged into an external 110-VAC power source.

Again, these Inverters, like the Converters in RVs, provide a limited amount of electrical power for the RV owner.

The RV owner should keep in mind that due to this current limitation with the RVs Inverters, which operate on the COACH batteries, should be turned OFF when not in use, although they are not designed to use very much of the RV's COACH battery power.

Why have GFCI Receptacles

First of all, GFCI is an acronym that stands for **Ground Fault Control Interrupter**.

GFCI receptacles are actually safety devices; they are special receptacles that are designed to "kick Off" if current flow is detected between the COMMON and GROUND wires of the GFCI receptacles.

The Ground Problem –

Electrical standards require that all Appliances manufactured in the USA not have the Ground and Common wires connected to each other.

The Hot and Common wires or appliances are not normally connected to any metal on the exterior of an appliance; while on the other hand, the Ground wire is typically connected to the appliance exterior shell if it is metal.

> **Tech Talk**: In a properly wired house, the Ground wire of your home electrical service is actually connected to Earth Ground normally near the entry point for the house wiring.
>
> In such a house, the person who handles a properly wired appliance will not be touching something with a voltage ON IT or receiving any voltage transferred to their body.

When this (or any) appliance is plugged into a GFCI receptacle, anyone who touches it will be protected from an electrical shock because the circuitry in a GFCI receptacle will detect any current flow between the Common and the

RV Electrical Systems

Ground wires and if even a small amount of current is detected, it will kick the GFCI's internal breaker.

RVs are wired to this same standard and also have GFCI circuits. Most home (and RV); kitchens, bathrooms and outdoor receptacles are wired together as part of special GFCI circuits.

Each of these circuits will have one, what I call **Master GFCI** receptacle, that can in turn, have from none to four or five **Slave GFCI** receptacles. These slave receptacles are wired to the GFCI Master receptacle so that if any current is detected between the Ground and Common wires of any of the receptacles in the circuit the Master GFCI will "kick Off".

GFCI Trouble-Shooting Tip: There can often be several such circuits in an RV and if any receptacle inside an RV suddenly loses power then the owner should always look to see if it is plugged into one of these GFCI circuits, and if it is, they should try to press the RESET button on the Master GFCI receptacle first.

Then, if it kicks OFF again, unplug all of the appliances plugged into the Slave receptacles and see if the Master receptacle resets then.

Cheater Power Cords and Home Receptacles

I should mention here that most of us RV owners will at one time or another want to plug our RV into a receptacle at our home, for one purpose or another.

To connect to our house, we will typically use our 30-Amp or 50-Amp electric service cable and then we will use "cheater adapter cords" to plug the RV into a receptacle on our house or garage.

When we do this, many RV owners will find that the house receptacle they need to use is a GFCI receptacle.

And often, once they get everything connected, they will find that their house GFCI receptacle will keep "kicking Off".

Well, here's the thing about GFCI receptacles; they kick out when there is only a very slight current detected between the Ground and Common wires.

But if you look at all of the cables and connectors they have hooked up, you will usually see a combination of dozens of extra feet of wires along with multiple connector points; all of which can be conducive to leakage currents.

> **Tech Talk**: Often, the combination of inductive coupled voltages between long parallel cables, and the typical buildup of oxidation and grime on the connectors of cables used outside will often add just enough current leakage between the Ground and Common wires for the GFCI to kick out on you.

So what's the solution? I can only suggest that if this happens to you, then you do have a good chance of reducing the current leakage if you use shorter cables and if you get those copper connections clean and looking like metal again.

RV Electrical Systems

COACH and ENGINE DC Power Systems

The typical motorhome will not only have a **COACH** (or **AUX**) battery system that provides 12-VDC power for the owner to use, it will also have an **ENGINE** (or **MAIN**) battery system that provides 12-VDC power for the engine and drive-train of the motorhome to operate.

A motorhome, just like your typical car or truck, operates on a 12-VDC electrical system, which includes all of the electrical parts that allows it to operate as a vehicle.

The owner of a gasoline powered motorhome will generally have a chassis and engine/drivetrain manufactured by one of the major truck companies such as; Chevrolet, Ford or Mercedes truck and the electrical system will be pretty standard for that manufacturer installed I them.

And this is also true of Diesel powered motorhomes, in that they have standard diesel truck-type engine and drivetrains in them, typically made by Caterpillar, Cummins, or Mercedes.

One good thing about this standardization of the 12-VDC power systems in a motorhome is that the RV owner can have almost every type of necessary service and repair performed on their motorhomes by the drive-train manufacturer's service centers across the country.

So, back to the two electrical systems in a motorhome, for the most part they operate almost totally separate from each other.

Refer to the diagram below labeled; **Simplified Generic RV Motorhome DC Power Systems dwg-05** to understand this special RV electrical connections.

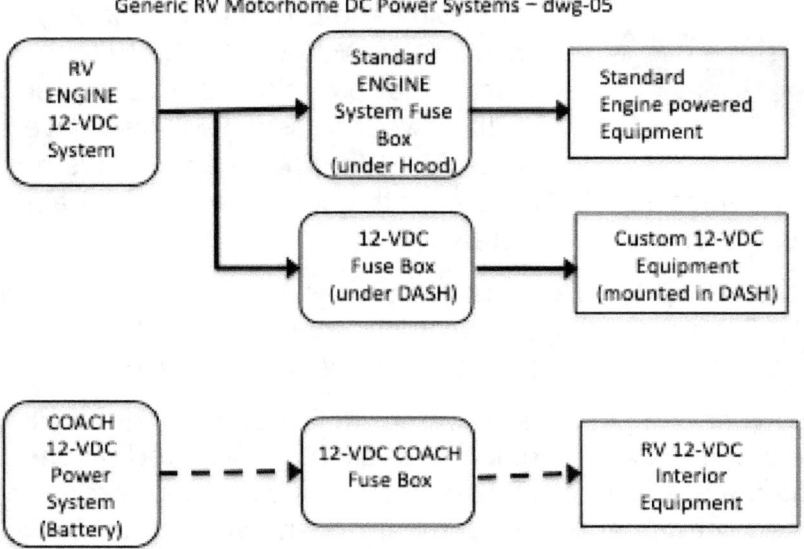

The CUT-OFF Switches

Probably the most forgotten and yet quite important pieces of equipment in your motorhome or other design of RV are the **CUT-OFF Switches**.

Because most RVs are only occasionally on the road, they end up quite often sitting in a campsite, or often seasonally in a storage site.

When an RV sits, even with everything turned OFF, the batteries will lose their charge.

Without going into the technical reasoning for this happening, to put it simply, the wiring in an RV is very long and when one of these wires has a voltage on it, it will "inductively couple" to a wire in the same harness of a lower potential and the source of the voltage will be discharged to it over time.

One way to prevent this from happening is to place a high current solenoid in series with the source line near the battery,

Then a cheaper switch can be placed somewhere convenient inside the RV that will open/close the high current solenoid contacts and thus remove all of the wiring in the RVs harness system from the battery.

When this is done the battery essentially has no load on it at all and will take much, much longer to eventually lose its charge.

By having a CUT-OFF Switch for the **MAIN (Engine)** battery and another for the **AUX (COACH)** batteries, the RV owner can go for weeks or longer at a time before they have to worry about having to recharge their RVs batteries.

Exterior Electrical Equipment

There happen to be several electrical items that although they are mounted on the exterior of a motorhome they are used most often when the motorhome is traveling.

Because of their location and the times during a trip when they are most needed, they are powered by the **MAIN** (or Engine) **electrical system** and not the **AUX** system.

These different pieces of RV equipment that are wired to the Main DC System are;

Entrance Door Light –

The light over the Entrance Door is needed the most often when the motorhome is sitting in a campsite, but it is often good to have it functioning when you pull into a campsite, gas station or Rest Area and because of this it is powered by the Engine battery.

Entrance Power Steps –

The Entrance power steps are necessarily retracted while traveling but they also need to be opened when the RV stops and the motorhome is in Park, for the convenience of the people in the motorhome to exit or enter.

The wiring for standard RV entrance steps includes a switch right inside the entrance door, a magnet activated switch on the entrance door itself, and if the Rv is a motorhome, there is a switch that detects if the transmission is in Park.

The way they are all wired is a little complex to explain here, but they are all there to make sure the RV owner can enter and exit the RV entrance safely using extended steps, while also making sure the steps are closed when the RV is traveling.

Also not that when the steps are opened the light under the steps is lit

Power Awning –

The main Powered Awning on a motorhome, like the door entrance light is most often used while the RV is in a campsite but it is usually powered by the Engine battery.

Power Leveling Jacks –

Leveling Jacks are common on motorhomes and numerous designs have been used on motorhomes as well as the larger travel trailers, over the past.

There are older RVs on the road today that use 3-point or four-point leveling systems that can be either hydraulic or electric.

RV Electrical Systems

The more modern systems are electric and they get their power from the COACH battery system the same as with travel trailers of all types.

A lot of RV owners will try to operate their Leveling Jacks and forget to perform certain steps that they should have done beforehand.

So you should always keep in mind that to operate your electric leveling jacks on a motorhome there are a few things that must be done properly;

The RV owner should make a quick check that the campsite is level enough for all of the jacks to touch the ground and then function within their range of movement.

- The transmission should be in PARK
- The Engine should be ON,
- The Parking Brake must be set.

Leveling Tip –

Most RV owners will carry several 12x12x1-inch blocks of wood to stick under any leveling jack that didn't reach the ground during the leveling process.

If the RV cannot be leveled because of one of the jacks; you can retract all of the jacks and place one or two of these boards under the jack base that is reaching its limit and then reset the Leveling System and try again.

Interior Equipment that operates on the MAIN (Engine) electrical system -

There are certain electrical devices inside a motorhome that, although they may be mounted inside the body of the RV they are also powered by the **MAIN** (or Engine) battery as they would be in that specific truck manufacturers vehicles.

Most of these devices can be found mounted on the dashboard of a motorhome;

- AM/FM/Satellite Radio, (often switchable between DC Voltage systems). Selectable with a dash switch for MAIN or AUX power.
- dash Heater/AC System,
- Front comfort Fans,
- 12-VDC outlets, on older RVs these are standard cigarette lighter sockets, on newer RVs these are usually USB connectors.
- Driver/Passenger reading lights,
- Rear and Side Camera System,
- Power/Heated Mirrors,
- As well as other unique equipment often used by the driver.

There will often be a separate fuse box for these devices in a motorhome, separate from the engine fuses under the hood, that will be located under the dash of the RV that provides fused power to these accessory devices.

The RV Propane System

Motorhomes as well as nearly all camper trailers will have a Propane tank System.

In a motorhome there is usually one large, 30-50-gallon propane tank, with its own meter and shut-off valve.

In other campers there will be one or two propane tanks whose size is dependent on the projected amount of propane that would normally be needed by that RVs devices that use propane.

There are several appliances in todays camper that operate using a combination of Propane, 110-VAC and 12-VDC and this section will describe how they are interconnected.

See the diagram **Typical RV PROPANE System Connections** – dwg-03.

2-Way Fridge

The standard 2-way Fridge used exclusively in nearly all RVs manufactured up until the last decade or so, can operate on either 110-VAC or on Propane.

This flexibility allows the RV owner to keep their fresh foods cold or frozen while sitting in a campsite and when traveling from one campsite to another.

But a lot of RV owners don't realize that the standard 2-Way Fridge has a Control Circuit board that actually controls the functionality of the 2-Way Fridge. It controls the inside temperature, the fan and whether the Fridge runs on 110-VAC or Propane.

This **Circuit board operates on the 12-VDC** from the COACH batteries and if they are not fully charged then the Fridge will not operate properly. If the battery voltage is low the Fridge will not operate at all.

On another note, be aware that the Ice Maker in a 2-Way Fridge operates only on 110-VAC, so always lift the water cut-off lever when you are driving down the road, or you could have a lot of water being pumped into your Ice tray until it overflows into the Fridge itself.

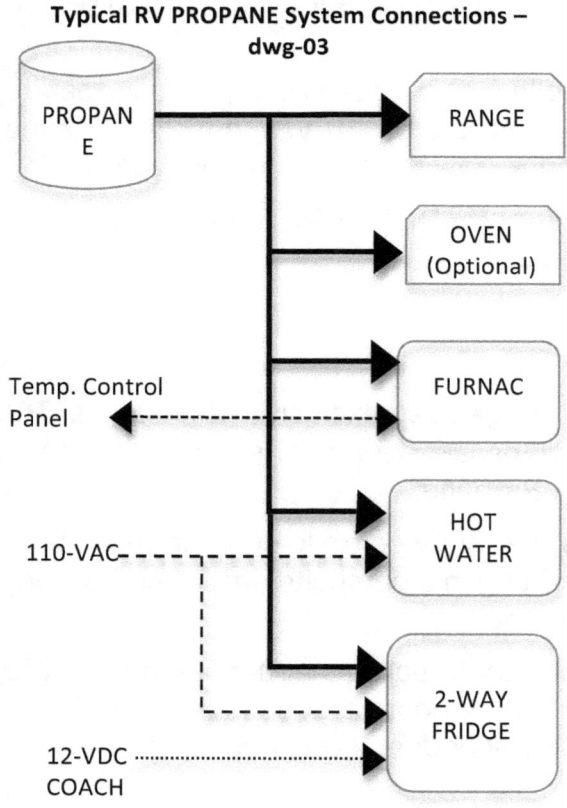

RV Electrical Systems

Propane Stove (Range) /Oven

You will find there are several appliances in RVs that are powered by the available Propane but be aware that some of these appliances also have Electrical inputs that must be there for the appliance to function properly.

Neither a propane Range nor a propane Oven require external DC or AC Voltage connections, and they are very easy to operate and maintain.

The only electrical power they will use is a spark from the built-in "Starter" on each burner.

Because the typical propane oven installed in most RVs is not very efficient or accurate in controlling the oven's temperature, most RVs today will not have a propane oven but instead they will have a very efficient Convection Microwave Oven.

With these new electric ovens you have the functionality of a home microwave oven, but you also have the capability to efficiently bake items; even cakes, pies and breads, using the Convection function.

Propane Water Heater

The Water Heater in an RV is designed to utilize either AC-Voltage or Propane to generate hot water.

You can select which way to heat the water with the switch that is usually found near the Water Heater itself; which is typically found under the Kitchen Sink.

Most Water Heaters will have a mechanical sensor on it to sense if the water heater tank has any water in it. If it is empty, then the Water Heater will not operate.

Propane Furnace

Probably the cheapest way to heat an RV in cold weather is to use a propane-powered furnace; and because of this nearly all RVs will have an efficient propane furnace in them.

Although the furnace itself does not need any external electric power, it along with the Air Conditioner, is controlled by the RVs temperature control panel (temperature, heat, cool, fan, etcetera), which in turn gets its power from the 12-VDC **COACH** battery power system.

A Home Fridge in an RV

So that they will be an economical addition to an RV, 2-Way Fridge units are, and always were, relatively small compared to the normal home Fridge that we all use.

And this size difference along with the customer preference for more food capacity, has driven many RV manufacturers to no longer install a 2-Way Fridge in their larger RVs.

Instead, they will often install standard home Fridge units in today's RV.

The problem with using a Home style Fridge unit in an RV was, how to keep the food cold for long periods of time in a fridge that only runs on 110-VAC while you are on the road or stopping for the night somewhere and do not have AC Power in your RV.

There were several potential solutions but it turned out that the most economically viable ones for the RV manufactures were relatively similar.

The first solution was to connect another Inverter to the COACH batteries and use it to provide the 110-VAC needed to power your Fridge while traveling.

RV Electrical Systems

The biggest problem with this solution was that the common house Fridge drew a lot of current and after 4-6 hours on the road, the standard COACH batteries were losing their charge from this extra load.

So, the first solution to this was to enlarge the COACH battery systems to include more batteries. Doing this gave the RV owner a few more hours of travel time and keeping their Food cold.

Some RV manufactures, rather than overload the COACH batteries, simply added a separate battery bank and Inverter just to power the House-style Fridge in the RV.

Both of these solutions worked well for most RV owners because honestly the majority of motorhome owners use campgrounds and are not "rough Campers" so they almost always had a power source to plug into in their next campground and recharge their COACH batteries each night.

But there were the "rough campers"; and they still wanted their RV to not need to be "plugged in" for days at a time, and their only solution was either utilizing a generator to keep their COACH batteries charged or own one of the many RVs on the market that have the versatile 2-Way Fridge unit in it.

Towing Connections

A book about RV Electrical Connections would not be complete without my trying to remove some of the confusion about the Electrical connections between a towing vehicle and a towed vehicle.

I think that at one time or another almost all RV owners have experiences a situation when they have a connection problem with the signal lights, or brake lights, etcetera; not working properly between their camper or motorhome or towing vehicle.

And once they reconnect or just wiggle the tow cable they find they have no idea how this towing connection system is wired.

Well, there are actually only three standard towing connection systems used with RVs and below you will see how each one is wired.

4-Pin Tow Connector

The standard 4-Pin Tow connector is probably the one found most often on trailers, and are designed to make the basic connections needed to pull a trailer and connect the necessary lights as can be seen below.

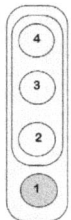

RV Electrical Systems

Pin #	Use	Wire Color
1	Ground	White
2	Tail lights, License Plate, Side Lights	Brown
3	Left Turns and Stop Lights	Yellow
4	Right Turn & Stop Lights	Green

6-Pin Tow Connector

A 6-Pin Towing Connector adds two more electrical connections that the 4-Pin doesn't have.

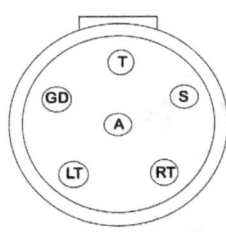

Check below for the details of the wiring of a 6-Pin connector

The **Electric Brake** pin provides a variable voltage to the electric brakes of large tow trailers and campers to help stop them.

Pin #	Function	Color
A (Center Pin)	12-Volts	Red of Black
TM (at the guide)	Tail Lights	Brown
GD	Ground	White
LT	Left Turn Signal	Yellow
RT	Right Turn Signal	Green
S	Electric Brake	Blue

7-Pin Tow Connector

The 7-Pin Towing Connector system is designed for and used most often when there is a heavy load to be towed.

See below for the connections and their purpose;

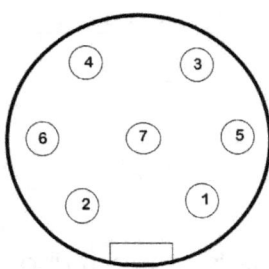

Pin #	Function	Wire Color
1	Ground	White
2	Electric Brakes	Blue
3	Tail Lights	Green
4	12-VDC	Black
5	Left Turn signal	Red
6	Right Turn Signal	Brown
7	Aux Power or Backup Lights	

RV Electrical Systems

Copyright © DONALD W. BOBBITT, All Rights Reserved

All Rights Reserved. Except as permitted by the U.S Copyright Act. No part of this publication may be reproduced, distributed or transmitted in any form, or by any other means or stored in a database or other retrieval system without the prior written permission of the Author.

Disclaimer

The information provided in this book is true and accurate as far as the Author is aware. Care should be taken to adequately research everything presented here, but if there are any errors present, the Author cannot be held responsible for the information presented nor any damage or problems incurred by or to the reader.

The Author assumes that the reader always follows proper safety procedures when operating or working on their RV and the Author cannot be held responsible in any way for any damage, injuries or problems encountered by the reader, under any circumstances.

Any Characters and Events presented in this book are fictitious. Any similarities to real persons, living or dead, is coincidental, and not intended by the Author.

THE END

www.ingramcontent.com/pod-product-compliance
Lightning Source LLC
Chambersburg PA
CBHW070818220526
45466CB00002B/701